AF247625

TIMES & LOCATIONS

MICHAEL SMITH

TIMES & LOCATIONS

THE DOLMEN PRESS

Set in Times Roman type and printed at the Dorset Press, Dublin,
for the Dolmen Press Limited, 8 Herbert Place, Dublin 2,
in the Republic of Ireland.

March 1972

Distributed outside Ireland, except in the U.S.A. and in Canada
by Oxford University Press.

SBN 85105 219 3

ACKNOWLEDGEMENTS

Some of the following poems have appeared in *Sumac* (Michigan),
Hierophant (Los Angeles), *The Denver Quarterly*, *The Lace
Curtain* (Dublin), *The Kilkenny Magazine* and *Hibernia* (Dublin).

Poems I-VIII are from *With The Woodnymphs* (New Writers'
Press, Dublin, 1968).

Poems IX-XXI are from *Dedications* (NWP, 1968).

Poems XXX and XXXVI were previously issued in pamphlet
form by Advent Books, London.

to two friends,
Paul O'Dwyer of New York
and Professor Lorna Reynolds;
and to the memory of my mother.

Locations and times—what is it in me that meets them all,
whenever and wherever, and makes me at home?
Forms, colours, densities, odours—what is it in me
that corresponds with them?

WALT WHITMAN

I

He fought death, it is true, to the very last
Cursing the white coldness creeping up his legs:
His birds were still singing in his brain.

His life was a mockery of two women's lives:
Marrying first Lust, then Pity, turning from her
To his yellow songbirds and his marvellous dream.

It was a note to be heard never yet heard,
A black tulip on the blue air—the unknowing
He lived among called it a pink elephant.

And there—where flies in summer darken the air,
Blackening the flypapers by the second,
In black gaslight and the pigs' smell and the poor's despair.

When the time came, I imagine the sun set beyond the high grey
    wall,
Beyond the innumerable railway lines and the Royal Canal,
And the birds at the foot of the bed were startled into song.

I I

The long lane of black timber huts
Desolate under the high windy pines
Where the wick burns into the encroaching dark . . .

The orderlies deal with our chaos there,
Efficiently,
With mop, bucket and syringe.

Flesh shrivels there, contracts and concentrates
To a final explosion of fluid,
The decrepit matter of the skull seeping away.

Before the end old men will not stay abed,
But try again and again to climb a non-existent stairway,
Failing again and again

Till the orderlies, familiar with these final antics,
Attach a bed-cage ; and age then is cotted in
For its unspoilt confining.

III

The butcher carried the cow's skinned head
By a bright grey hook in its eye:
Blood trickled on the dirty pavement.

The red-faced butcher in the blood-stained
Apron felt only the awkward weight
That pulled on the bright grey hook.

And there was the sun, too,
A hot noon sun; and children passing
Who said to their mothers

" Mama, Mama, look at the moo cow."
But their mothers dragged them by,
Past the cow's skinned head and its hooked eye.

I V

The great snake with dragon head
Climbs upward, coil after terrible coil.

It climbs upward behind the naked woman
Who stands, like some public monument,

In the dead centre of a single coil—
Her hair is tawny as the great snake's scales.

Behind a timid waterfall the evening sky
Is a deep blue pool, hoarding no stars.

Still, the moon is there—dull as a brown mud—
And a few clouds like small white worms.

Beside the snake and the wide-eyed woman
The prostrate sleeping man awaits his genetrix.

V

for Jack Sweeney

Somewhere in Clare beyond fields and rutted laneways
Two herons lifted into the grey air.

I climbed a promontoried Norman keep
Whence eyes swam over seas of grey rock ;

And the accompanying dog at the head of the stair
Trembled and whined in a grey fear.

The orchard was deadly sick with a grey fungus
And the bark of the hazel was silver grey.

The hair, skin, eyes, teeth of the old woman
Who polished my shoes for the dance, were grey.

Here in October dawn breaks in sheets of grey glass.

V I

Long Lane, one of Dublin's back-streets,
Just a street as Mangan struggled down,
The winter darkness and the pitchy dampness . . .

Then, like a horrible figure out of mythology—
A sphinx—the silhouette of this club-footed man
Against the glaring yellow of the lamplight.

As the branch tip tapping on the window glass
In the snow-stormed night becomes the lost object,
To return always when a sense of loss occurs,

So fear comes now with twisting twisted legs,
Hobbling down Long Lane a winter's night,
Climbing steps, then knocking at a door, and passing in.

VII

Since now you're dead and I, I saw you dying
and hours before the death
when life was flickering
delicately
like a minute shadow . . .

Through the window of your cubicle I saw
the first fall of snow
falling
falling
delicately
as only snow can fall

From long thin grey clouds like distant trains
travelling across the sky above the pine tops
their destination inscrutable as your silence

As you lay upon your final winter bed
your face shrunken black upon the pillow . . .

Silence.
And then . . . *A stairway! A stairway!*
At the end of a corridor to nowhere
A stairway!

While the first snowflakes disappeared
into the black square of the chest hospital
A stairway! A stairway!

VIII

Because he'd lost his nerve,
    when they threw him out
he became a bird,
    growing beautiful feathers,
flying from hillock to hillock,
    feeding on delicious watercress
and, at night,
    perched high up in a tree,
dreaming of the wife
    he'd left behind.

Once, for some inexplicable reason,
    he was readmitted
to polite society,
    meaning human society,
but an old leering hag
    starting him off again
with a ridiculous game of leap-frog;
    the feathers grew
and, airbound once more,
    there was no going back.

Finally, despairing of rehabilitation,
    he began to like being a bird,
to enjoy the watercress
    and the heady delights of flight
over summer landscapes.
    In winters of black frost,
with big brown wolves howling
    beneath the tree he's perched on,
he waits, in silence, for the dawn,
    with passionate, burning-red eyes.

I X

The desolate rhythm of dying recurs,
The rhythm of outgoing tides, corrosion of stone,

Fall of petal, of leaf and soft rain on empty squares,
The fading memory of song, say, in an old man's head

That never stops in a moment of time,
A rainbowed vertigo spinning beyond the nurse's cool hand,

Subsidence of wind and branches against a settling sky,
And stars fading at dawn, or fall of snow:

Something ordered, yet desperate and violent—
A rose, say, or an old man's humiliation.

X

Dead days, like a vegetable world all burnt up:
Blossom and weed, rose and pimpernel, gone to the flame's mouth.

And there is no point in asking, no one can tell you why,
There is no reason for dying, it just happens that way.

In a cobbled lane of stinking pig and lesser celandine
An old woman emerges from the cave of an upturned car, her home.

With querulous voice she hails to every passer-by she sees,
Pointing to the green church dome among the hawthorn trees.

To every passer-by who will not mind her
The moon is a greener cheese, she says, *only kinder.*

X I

Through no imperial portals, but rusty bars on broken hinges,
To this kingdom of black earth like dampened dust

Where my green knight evades the black-shawled witch's eye,
The dragon's teeth and the sly pervasive worm.

Bells beyond the kingdom toll the significant hour
And the streets are silent, the squares empty.

In his gaslit room of the golden birds
Roosting quiet as the small rain of summer

The old man receives the boy's green gift,
And the green knight hears his golden song in wonder.

XII

There is something white and still in the black river
      below the cathedral—
In the black river quiet as the dark it flows through—

And no one can see it but the still man across the parapet.

Streets have fallen taking pieces of the sky with them ;
Summers have passed taking many who walked through them ;
The spring is busy computing a new mortality list.

It has rained,
            it has snowed,
                        and even the sun shone
Occasionally.

But the gaze of the still man across the parapet
Is fixed on something white and still in the black river.

XIII

All things would speak to me, stone and wood,
A child's cry, odours, and flowers, especially flowers.

And what could I then say to these echoings—
Forever locked in my anthropomorphic lunacy.

Shine, moon, on no catastrophe—the spinning wheel,
The bearded mouth's spasmodic ooze of blood;

Burn on, sun, and keep the maggots warm
For the infant at the breast, a mother wholly gone.

And I would say, in my anthropomorphic lunacy:
Heave ho for ay, you salamander heart.

XIV

Too soon put up for the wind that blew it down,
    Hope became despair,
And despair was the sea on which my ship had sailed.

The rain became the city, the city became the rain,
And the unnested fledgling that ancient playful dog.

Barbara, Barbara, my canal-bank pinkeen girl.
Your curls are in the water with the barge,

And I am sailing down a purple Nile
Across the mill's aluminium-silver dome

From where the knacker's leans against the sky
And a solitary lilac weeps in its concrete yard.

XV

Here is the abattoir where, in the old days, were heard
The ultimate cries of table beasts.

Here is the home for unfortunate women
Who launder a twelve hour day and are never cleansed.

Here is the Union, picturesque on a sunny Sunday,
In fact, stables of appalling decrepitude.

Here are the slums where life swings back and forth
With a thud like a heavy pendulum.

Here is, partly, love's ecology: occasional blue skies
And, more often, thunderous falls of black stars.

XVI

There is one, never seen, behind that window
Where the light burns on through every night ;

Another, whose sole appearance in a generation
Was a naked dance with obscene gestures.

There is a third, whose tired white face
Forever droops from a backyard window

Like some old flower that will not wilt
Administered by an ancient spinster :

Mad but quiet people, gone, like Elijah, up,
Body and soul, into their their own home-made heavens.

XVII

It was, the neighbours say, a considerable time ago,
But nobody knows the date or exactly what took place.

It was late summer, say, just for the neighbours' sake ;
Imagine the afternoon, the flies maddened with heat,

The flies thick in the air, exploring every place,
And the afternoon there in a deadly prurient heat.

Even the old man's pigs lay prostrate in the square
Though grunting occasionally at a black fist of flies.

It was late summer, say, near the seasonal change :
But nobody knows the date or exactly what took place.

XVIII

Where the copper grass rises high as the sky-roof
From its so small earth in the crumbling masonry ;

Where laughter collides in the incongruous apple tree
With this now all too inclusive sunlight.

Cobble, canal, eccentric wagtail,
The indescribable bundle of the vagrant's belongings,

Caught and included now, feather, stone, water, flesh :
And O my child, and my friend, and my love,

Where the copper grass rises high as the sky-roof
From its so small earth in the cracking joints.

XIX

You arrange and rearrange the furniture,
You water the flowers
And straighten the old tired pictures
Done by a hobbying grandfather.

Everything looks fine for the visitor.

You exit and a moment later re-enter.

Yes, everything looks fine,
And even the old sun is shining for you
Through grandmother's lace curtains.

X X

The linnet on the orchard tree was green
So the day became an emerald, the branched moon a pearl.

Menacing gargoyles laughed below the turrets
But poet and etcher were too tired to hear.

And the world was tired of their excessive love,
Unsympathetic now to all their fumbling gestures.

The stars were inedible, the clouds could not be caught,
The beloved married another and went away:

Arcadian lovers, dropt from the silver lining of their dream,
Their legs danced in a last impatience with the world's height.

X X I

Ghost, come and speak with me beyond the hawk's-beard,
The scum-surrounded yellow water lily ;

Tread gently the warm crushed brick of home,
And, upon the thistled hill of childhood,

Speak to the assembled sunlit dust,
The heavenly host in all its summer heat.

Afterwards, in needless demonstration,
Walk the canal waters' hot and putrid surface.

I, like the dog's black skull in submarine sleep,
Await your oracles from beyond the rubbish heap.

XXII

for Kay

Beneath the table the naked boards,
Three feet of air above the black-earth kingdom
      of the singing worms,
And the familiar trapdoor unbolts to the post-midnight,,
The single shaft of moonlight through the curtains
And the boom of the clock with the tiny pendulum.

Once more the gnomic face with the huge fish-like eyes
Glares through the opaque darkness ;
Its breath is heavy upon the air like a living thing
Fouling the fevered night.
A moth flutters against the window and is still.

Quietly the memory wheels to the past like a bird of prey.

XXIII

The church steeple fingers the sky.
The cold air glints like frost against the stars.
And the moon's casual indifferent face
Simply there in the broken sky,
Bird-high above the small terraced houses
Where people dream in the hard night
Of their own private suns:
Red things beating still beneath the packed snow.

XXIV

Here the walls breathe to the imagination's warm reception.

Shadows are everywhere, real and unreal.

Questions form in the darkness and remain there.

Glistening machines throb in antique skeletons.

A tree is an incongruity where bricks are growth.

The canal stagnates on others' putrefaction.

The convent's bell is synchronic with the horse's last neigh.

Over the vendor's shoulder the hungry cry.

Poems for souls lost between hand and mouth.

X X V

Finding no ghosts we must invent our own:
Ghosts that will admit us to our own importance.
It is unfortunate our material lacks variety.
People will tell us we are morbid.

Canals and old houses, preferably uninhabited,
Weeds and common birds and marketplaces,
Sunlight down laneways making odd pictures,
Imposing green domes above such squalid life.

And we walk for an evening when the air is cool,
Past the de luxe hotel like a de luxe chocolate box.
We observe people as though we were not one of them.
We agree love is not something you find on the streets.

As usual we make for the canal as something clean ;
We would, of course, prefer a mountain freshet.
Sometimes we are sick of our excremental piety.
We agree we should go elsewhere, when the time comes.

XXVI

The long Wednesday-dunged road from the cattleyard to the bridge
Downhill past Summerhill to the Old Maids'
Where six summered trees on high soil occlude the tenements
Where Tessy lived in the short short high day of the vacant child.

The hall is dark yet bleak with the bare brick of stript wall . . .
O old husbandless woman of many children all your own,
Shawled in the shelling twilight of curtained light,
The sun a great sheet of light outside the narrow windows,

True quiescent love, mother to us all,
Unknown to us all, and unknowable,
There in the black fathoms of life itself,
Lady of all life, beautiful and sordid.

Wordless love, queen of the living and dead,
Radiance through wrinkles, only true mirror,
Principled silent lady of the bed, the heart-breaking mirror, coffin,
Be with me still in my last loneliness.

XXVII

The time was truly right for the recall of stories.

The flesh withering to its own painful tune,
The memory tightened its cords
For its last intense melodies ;

Love was crying out beyond the slow sad death,
The collapse of things like the sift of sand.

XXVIII

for Peadar

The dark philosopher speaks to his congregation
In no sunlit groves of academy ;
The moon floats slowly across the evening sky,
A severed petal on the canal waters.

' O melancholy brothers ! '
The voice rises from the clammy reeds,
An exhalation of gas configuring in the still air.
' Now, at last, the word ! '

Sighs sick with grief
Scrape softly on the lichened granite
Pleading.

' Stars that wheeled to the sad music of our birth
Gyrate now to our cacaphonous death.'

The voice rises higher still in the breathless night.
A dark bundle slips from above across the parapet.
It shatters the night's glassy surface ;
But the hearers hear only the slow sad voice.

' The eye does not mirror but distorts ;
It does not wish to see what it is that hurts.
Lights in the windows on a winter's night
Consummate the vagrant's fumbling dream.

' Remember—
Childhood's always golden summers
When sunlight flamed in the air a huge impassioned deity,
Recumbent even on the piggeries' squalid stench :
And the child's feet carried him everywhere to wonder.

' Everything was special and you made the laws
Of like and dislike, can and cannot:
The wagtails came when you called
And the mongrel's scent was love.

' But childhood's summer wheels away from wonder.
Its fruit is unpluckable and when it falls it rots:
The young girl with the gold-shaking tresses
Sings no more down scented lanes to the vagrant crone
In the alley's cavernous mouth.'

The voice halts now, and then restarts
To the shuffle of feet and broken glass:

' My brother, my poor brother, it is thus:
Boats gliding like dark shadows of a dream.'

XXIX

When he died it was still summer.

The street rose and responded with its rehearsed platitudes,
Innocently, knowing no other way
To remark his casual but final disappearance.

The girls from the sewing factory
Passed by at the end of the street
Singing as usual that Friday evening.

It was still summer, and like a heavy gauze
Sunlight caught in the stench from the abattoir;
But no one remarked this olfactory unpleasantness.

His children played in the street that evening
Working out their own private poems on bicycles
Or dancing gaily on the pavements
To the frenzied buzz of the blue-bottle against the window glass.

X X X

for James Liddy

Blonde hair at the edge of the pavement
caught in the sunlight like a bright leaf:
the dull classroom as far away as the heaven
of angels' white wings that never crumple.

The lorries passing, the pigeons, the gulls;
the smells from the factories, soap, chocolate, timber;
and the bandy drover's shouts that rise forever
over the thousand brindled backs of the cattle . . .

How infinitely important all that was
to the wide eyes under the blonde hair.

A big stick poised to beat their backs
when the drovers turned theirs . . .

What sense of importance swelled at the smack
of the stick on the dung-caked dumb flesh.

X X X I

for Gerard Smyth

Always the old houses were falling into dust ;
and the weeds grew high as the houses fell,
and gaps in their broken walls led to strange places.

Piles of red dust burned in the sunlight.
The slow, blue pigeons. The purple weeds.
A girl's soft laugh rounded the broken wall ;
golden shards, old china, among the nettles.

Black, that city soil was black when you plucked
grass sods to drown the drumming on the tin roof
of the rain that always fell at the close of evening
as you listened quietly to the tense growth of sounds.

It was almost enough to have stayed there forever,
or at least till the winter came and all was frozen.

XXXII

At the appointed hour they came
from who knows where, or who knew then ;
from the farthest reaches of the city,
from park bench and slobland and river's edge ;
from hostelries and distant places beyond time.

I can remember no details to speak of.
A queue of greasy men against the railings
moving to their own slow tempo
or the whiskered nun's brisk propriety.
I can recall the hour from the tolling of bells.

In the bushes the caterpillars ate voraciously
at the sooty leaves.
The cakeman came with trays of cakes
caught in their last act of decay,
dessert and industrial refuse.

I remember the summer as hot pavements.
The time is imprecise, but the place is there,
and the wailing of factory sirens,
and the workers' clean sweep on foot or bicycle.
In the evening the air sank with the weight of darkness.

XXXIII

An instant comic opera:
the singer's voice howling
to amuse the balconied crowd
while thirst tore at his throat.

They appeared weekly,
young and old, always male;
at teatime on Friday when coppers
were plentiful as stomached pints.

The children were most amused,
the scene's rich comedy rising for them
higher than the soaring voice;
its lapses, gay winter slides.

XXXIV

for Burton and Peggy Feldman

It was the time of the big races
when prize pigeons, lost, walked into your hands.
We got up early and waited patiently,
watching the sky or, for diversion,
the sly cock's annoyance of the silly hens.

The afternoon stretched itself like a tired walker,
with the pigs and eternally old women in doorways.
My grandfather's room of singing birds,
canaries, finches, sang through the open window,
conjuring trees and streams to shut eyes.

At three in the afternoon,
in search of a cup of sugar or spoon of tea maybe,
a neighbour stumbled to a horrible difference.
She neither fainted nor screamed and,
by four, even the murderer was arrested.

It was the time of the big races . . .
But something broke, anticipating
the blue hatchet's thrust to the skull.
Red blossoms of geranium
sprinkled the sunlit yard.

X X X V

It was a tropical summer
even with snow falling past the open window,
and the air inside was thick with dust
even as the grey air outside with falling snow.

You gave no thought to that, old man.
Death was merely what happened to friends,
how they took leave of you ; and, as you reasoned,
there must always be a departure.

I was a boy, the factory a rough world
that made growing up even more difficult.
You were no philosopher to my ills
but a sad man saying " Yes, I know," making the balm.

The furnace roars still into the night.
The sirens wail-in the winter evening.
This is an elegy for no satanic stoker
but an old man who fell over the edge and was missed.

X X X V I

for Brian Coffey

HE:　These are dark things to speak of in a town like this—
　　　not from any ponderous, sentimental guilt,
　　　an imposed disposition to weep.
　　　It is so because it was dark and night.

SHE:　An inner and outer darkness of night
　　　that would grow naturally in time to light
　　　but could not be forced, not even slightly.
　　　Something was burning through to put things right.

HE:　As a child I danced my mad dance of sex.
　　　I was, in turn and together, pimp, prostitute, lover,
　　　belly dancer and passionate sodomite. I prayed,
　　　both monk and frightened sinner in a darkened box.

SHE:　Nothing was itself but the burning flesh
　　　consumed what it touched in a dead heat of lust,
　　　throwing not flames but dense shadows of smoke.
　　　A limp couch was flesh I gladly lay upon.

HE:　Dark things to speak of that linger yet,
　　　follow me down these streets like sorrow's child.
　　　Time that was surely more than I can guess
　　　is quietly lost in a warm ingenious night.

XXXVII

When the sun debouched into the child's empty summer
Bringing with it
                white butterflies
                flying lost
                in a glowing world
                of brick and concrete
Or
                in the evenings the sudden
                frightening beauty
                of a tiger moth
                in a dark and dusty corner

It was then she took us,
The others too busy
Or worn-out with survival,
Or afraid of water,
To the sea
        beyond railway bridges
        and the din of industry
Where all the day through was filled with
                            sand and spray
                            and a wind
                            almost visible
                            rolling in bulks
                            across the sea's surface.
And there were
                kisses
                and touchings
                in the spiky grass
                amid the dunes.

Then the journey home in the train ;
Even that, a departure, was not sad
Because of her.

And the walk down by the long high wall
With buckets of periwinkles and crabs.

I recall the slow furniture of her swollen feet.

XXXVIII

The retired soldier spick-and-span in every detail
as his beautifully groomed silver moustache,
how appropriate this his last duty,
sentinel at the silent station of the city morgue.

No grieved relative turns to him for compassion.
He is no more to them than the cream, aseptic walls,
the tressels or the green-veined marble slabs,
part of a scene to be confronted and forgotten.

To the not-too-stricken he will comment on the weather,
an inclement season's frequency of death,
the futility of debts incurred by funerals ;
and then go to his lunch of chops and liver.

Once a body stirred, he recalls.
Bravely, he waited for further signs of life ;
but the stirring stopped beneath the sheets.
Crossing himself, he closed the window on the summer wind.

XXXIX

Small things, like the turning of a key,
open, as light does a knot of petals,
the mind locked lost in a dead routine.

Things small enough to have seeped through the pores,
the nerves, the blood, into the brain,
till they are forgotten, if ever remembered,
                and you, and me.

You smile. You clutch my hand.
Lovingly and sadly I feel the hard skin of the palm,
and its warmth touches me like a whole body.

Such incredible foliage outside the window.
The sun blazing on its myriad greens.
And the woman sitting at the bed beside,
a stupid lump of her neck, big as a fist,
tensely counting the puffs at her cigarette—
one—two—three
and the white blue smoke fading
                gently across the glass partition.

Teatime. We go outside into the sunlight
and wait. We remark the lawn's green,
the empty pond and the heat
falling in thick waves from the brilliant sky.

A steeple clock chimed.
Its leisured sounds lifted ponderously
like great birds:
but we were too frightened to laugh.

X L

for Irene

Now the night prowls round my heart
like a great feline:
I cannot tell fear from loneliness.

The air thickens with ghosts familiar and strange.
Your father's footstep creaks the loose stairboard.
A picture suddenly tilts on the wall.

Our human losses of these past two years
have left their wrecks
whose bare ribs protrude in this low tide.

A prayer comes to my lips from childhood
but cannot be said:
our lost mariners are unreported yet.

I say the word love into the mocking stillness
of the night that I'm afraid of
only less than losing you.

X L I

for Trevor Joyce

The great white horse my father drove
Hoofed up sparks at the closed door
Till it opened and my mother broke
To sympathy with the dumb beast in the hard weather,
Fed him more bread than we could well afford.

All that winter, every day, my father says,
 The great horse pulled at the reins as he passed
Our street. A beast easily harnessed to habit,
He had to be given a slack rein.
It was a bad winter and even the snow came.

Just after Christmas he slipped on the ice,
His great weight removed from the stability of earth.
His fall broke the two shafts, my father says.
His frantic breath clouded in the cold air
Till the vet shot him dead there in the street.

The story goes . . . A great white horse,
A huge gentle creature, a mere dray nag,
Slipped on the ice and broke his bones.
Time was then for the listening boy the unmeasured
Heartbeats, black hammers in the melting snow.

XLII

Who will speak now or ever for Suzie
Whose confinement has stretched for ten years
And will doubtlessly stretch for longer
Till her frail body, wracked by a straying head, breaks
Like the suds on the tub she bends over?

Suzie and the old woman of the Coombe
Who bares her crotch to the startled respectable
And screams ruin on all exploiters of the cunt.
Suzie in her laundry convent tortured with salvation,
And the old woman in her back lane chiding the crawling dark.

And their daughters, products of aberration, themselves aberrant,
Lost in the city's asylums, in their delicate heads
Wandering an irregular cell's sly curvature.
Not even filed as missing persons, it is as though
They didn't exist, hadn't happened, like last year's pain.

XLIII

for Ken

The dream
life's flotsam
sprouts always the gayest flowers

And if impossible
why life's and death's
as necessary, easily

And everything
sooner or later
falls short, even to the expectant.

You, too, have seen
the dust of blood slip
from the tired bricks of the old house.

XLIV

for Liam

All winter the old man slept
on the pavement under his tent of rags,
too mysterious for pity.

His eyes stared straight ahead
at the facing brick wall,
as impenetrable.

Down the street the hostel doors
had closed on his long winter.
Nobody cared.

He is so alone
he must be dreaming
of the soft density of meadow grass.

X L V

for Kay

We have seen better days
like old soldiers

We were complete
in the blood's unawareness

No tomorrows lurked
around any corners

The old woman's feet
were cold for your warmth

You are the stillness
in the ticking moment

XLVI

for Barbara and Lorna

Tumultuous rainfall of birdsong
at this the door of another summer,
through a slow, insidious dawn,
steel-grey, like an indefinite sword.

Sleep, children, out of harm's way,
till sunlight drives the ghosts
from the swaying air.